Logic

Third Edition

STEVEN ROMAN

Emeritus Professor
Department of Mathematics
California State University, Fullerton

Logic
ISBN 1-878015-19-2

Preface

This is one in a series of mathematics modules designed for the general college level student, whose background may only include intermediate level algebra. It can be used, together with other modules, or as a supplement to another text, in courses such as liberal arts mathematics, finite mathematics, or discrete mathematics.

The first chapter of this module is devoted to the most elementary aspects of logic. In particular, we discuss how to translate sentences into symbolic form, and how to construct truth tables. The third section of Chapter 1 deals with the concept of logical equivalence, where we emphasize DeMorgan's Laws and the relationships between a statement and its converse, inverse and contrapositive. These examples are especially important, since they illustrate how we often misinterpret the meaning of relatively simple sentences in the English language.

Chapter 2 is devoted to two applications of elementary logic. First, we discuss how to determine whether an argument is logically valid. Then, we discuss how logic can be used to design simple circuits, including a circuit for controlling an overhead light using three independent switches.

Chapter 1 can be used by itself to provide a brief introduction to the subject (about 5−6 lectures.) The entire module should provide about 7−10 lectures worth of material.

Answers to the odd numbered exercises appear at the back of the module, and answers to all exercises are available to the instructor upon request.

Preface to the Series

This series is designed to provide textbook material for a course in contemporary mathematics for college level students. For one reason or another, large publishers have not responded to the educational concerns of instructors. Through the use of desktop publishing techniques, this series of modules can provide the needed flexibility to adapt to the differing concerns of instructors and motivational needs of students. In particular, by using these modules, instructors can now select topics on a class-by-class basis. We hope that this series will provide both students and instructors with an adaptable learning tool to help increase enthusiasm for mathematics in the classroom.

Acknowledgments

I would like to express my indebtedness to Professor John G. Pierce for making a number of constructive suggestions for improvements in this module. Also, I am indebted to Ms. Donna Dolan and Ms. Joan Sholars for carefully proofreading the module and working all of the exercises.

Contents

Introduction

About 350 years before the birth of Christ, the Greek philosopher Aristotle began what we might call the modern discipline of logic. Aristotle's primary goal was to set down rules by which it would be possible to tell whether or not a given argument was valid, and this will be one of our goals as well.

Surprisingly, however, modern logic can be used in a way that Aristotle could never have even dreamed of, namely, it can be used to help design circuits for computers. Another one of our goals will be to see how this can be done.

Even though we will not be concerned with the history of logic in this module, it might be a good idea to briefly discuss the development of this area of mathematics, since it will give us an opportunity to describe what mathematicians do for a living.

While the logic of Aristotle's time could not solve all logical problems, it did a fairly good job until about the nineteenth century, when some rather profound mathematical discoveries took place that shook many people's confidence in the consistency of mathematics.

Bertrand Russell
(1872-1970)

In particular, mathematicians of the nineteenth century (and earlier) discovered several *paradoxes*, that is, statements that seem to contradict themselves. Perhaps the most famous paradox is the one discovered in 1901 by the philosopher and mathematician Bertrand Russell. A popular version of *Russell's paradox* goes as follows.

A certain town has only one barber, named Jack. Jack has a rule that he shaves those people, *and only those people*, who do not shave themselves. But then who shaves Jack? If Jack were to shave himself, then according to his rule, he could not shave himself. On the other hand, if he doesn't shave himself, then according to his rule, he must shave himself. Thus, Jack must both shave himself and *not* shave himself! This is a logical contradiction.

As you might imagine, the discovery of this, and other, paradoxes caused the mathematicians and philosophers of the nineteenth and early twentieth century to question the very foundations of their disciplines. Even to this day, we do not have entirely satisfactory ways to resolve the paradoxes.

This led, at the turn of the century, to an effort on the part of many of the greatest minds of the time to go back to first principles, in other words, to start all over again, and attempt to place mathematics on a solid foundation, free from all possible contradictions, such as Russell's paradox.

The new logic that came about as a result is usually referred to as *mathematical logic* to distinguish it from the *traditional logic* of Aristotle. Mathematical logic is founded on what logicians call the *formal axiomatic method* which, roughly speaking, means two things. To develop a mathematical theory, such as arithmetic for example, we must first set down in a precise way those facts, called *axioms*, whose truth is to be accepted without question. The intent is that these axioms should be so "obvious" as to be self-evident. For example, if we were to axiomatize arithmetic, we would want one of our axioms to be the statement that,

$$x + 0 = x$$

for all numbers x. Another axiom of arithmetic is

$$x + y = y + x$$

for all numbers x and y. (You may know this as the *commutative law of addition*.)

The second thing we must do in the axiomatic method is to set down the so-called *rules of inference*. These are rules that enable us to deduce additional facts (called *theorems*) from the axioms. Of course, once a statement has been determined to be a theorem, we can also use it to help deduce other theorems, using the rules of inference. Such a deduction is referred to as a *proof*. Thus, by means of proofs, we can build up a collection of facts (theorems) from the axioms of a mathematical theory.

In fact, this is precisely what modern day mathematicians do. That is, they attempt to discover new theorems (and their proofs) in the areas of mathematics that interest them (for example, set theory, logic, algebra, geometry, probability, statistics and so on.) In some cases, a mathematician may attempt to solve a famous unsolved problem, which is nothing more than a statement that many others have tried to show is a theorem (by finding a proof for it), but without success.

Thus, according to the axiomatic method, a statement is not a theorem unless it has a proof, and this is why mathematicians seem to have such a "thing" about proofs. It is only when a proof has been carefully written down does a mathematician know that his or her theorem follows logically from the axioms of the theory.

You can see from this discussion that the field of logic plays a very important role for those mathematicians and philosophers who are concerned with the foundations of mathematics. However, elementary logic can also benefit us by helping us understand the meaning of sentences that we use in our everyday lives. It is surprising how often we misinterpret the meaning of even relatively simple sentences. For instance, consider the sentence

$$\text{It is not true that I have a cat and I have a dog.} \qquad (1)$$

This statement has the same logical meaning as *one* of the following sentences.

$$\text{I do not have a cat and I do not have a dog.} \qquad (2)$$

$$\text{I do not have a cat or I do not have a dog.} \qquad (3)$$

Do you know which one?

To answer this question, we use a method discovered by Aristotle. Aristotle had the idea that the logical meaning of a sentence could be made clearer by putting the sentence into *symbolic form*, that is, by replacing portions of the sentence by symbols, or variables.

In our case, we let the variable p stand for the statement "I have a cat" and q stand for the statement "I have a dog". Then sentences (1), (2) and (3) can be written in the *symbolic forms*

$$\text{not } (p \text{ and } q) \qquad (1')$$

$$(\text{not } p) \text{ and } (\text{not } q) \qquad (2')$$

$$(\text{not } p) \text{ or } (\text{not } q) \qquad (3')$$

respectively.

Now, we will learn in Section 1.3 that statements $(1')$ and $(3')$ are *logically equivalent*, that is, they have the same logical meaning. (This fact is referred to as *DeMorgan's law*.) However, statements $(1')$ and $(2')$ are *not* logically equivalent, that is, they do not have the same logical meaning. Hence, statement (1) has the same meaning as statement (3)! (Was that your answer?)

Our plan in Chapter 1 is to set down the basic principles of logic. In particular, we will discuss how to put statements into symbolic form, and what we can learn by doing this.

In Chapter 2, we consider two applications of elementary logic. First, we discuss how to determine when an argument is valid. For example, the argument

$$(4) \qquad \begin{array}{c} \text{If today is Friday, then I must go to the bank.} \\ \text{Today is Friday.} \\ \text{Therefore, I must go to the bank.} \end{array}$$

is valid, whereas the argument

If today is Friday, then I must go to the bank.
I must go to the bank.
Therefore, today is Friday.

is not valid.

As with individual sentences, Aristotle knew that the meaning of an argument could be made more apparent by putting it into symbolic form. In this case, if we let the variable p stand for the phrase "today is Friday", and the variable q stand for the phrase "I must go to the bank", then argument (4) can be written in the symbolic form

If p then q

(5) p

Therefore, q

Aristotle showed, as we shall do, that any argument of the form (5) is valid, *regardless* of what phrases are represented by the variables p and q. (It is interesting to note that the people of Aristotle's day considered the ability to argue persuasively and validly to be a high art form. In fact, in those days groups of teachers, referred to as sophists, would travel like wandering minstrels, teaching students for a fee how to argue logically on a vast variety of subjects, such as mathematics, natural science, philosophy and law. The sophists would also engage in public debates, often commanding large fees for their appearances.)

Our second goal in Chapter 2 is to see how logic can be used to construct circuits for computers. (Don't worry — we won't assume that you have a knowledge of computers.)

Chapter 1
Elementary Logic

1.1 Symbolic Form

Our first goal in studying logic is to learn how to translate sentences from ordinary English into symbolic form. For this, we require some terminology.

A **statement** is a sentence that has a truth value, that is, a sentence that is either true or false. For instance, the following sentences are statements.

1)   All birds are animals.
2)   Today is Tuesday.
3) $2 + 2 = 4$
4)   If I study hard, then I will get a good grade in this class.

On the other hand, the following sentences are not statements, because they do not have a truth value.

5)   No smoking.
6)   Do you have a red balloon?

Statements are generally grouped into two types — *simple* and *compound*. A compound statement is made by combining simple statements using special words or phrases, known as **connectives**. The five basic connectives that we will study are

and

or

not

if...then...

if and only if

These connectives are listed in Table 1, along with their names and special symbols.

For instance, the statements "It is cloudy" and "It is raining" are simple statements, but the statements

It is cloudy *and* it is raining.
It is *not* raining *or* it is cloudy.

and

If it is raining, *then* it is cloudy.

are compound statements. (The connectives have been set in italics for clarity only.)

Table 1-The Connectives		
CONNECTIVE	SYMBOL	NAME
and	$\wedge$	conjunction
or	$\vee$	disjunction
not	$\sim$	negation
if...then...	$\rightarrow$	conditional
if and only if	$\leftrightarrow$	biconditional

Let us consider some examples of statements and their negations.

Example 1

STATEMENT	NEGATION
Mary is a good student.	Mary is not a good student.
$2 + 3 = 5$	$2 + 3 \neq 5$
Some cats have fleas.	No cats have fleas.
All dogs have fleas.	Not all dogs have fleas, *or*
	Some dogs do not have fleas, *or*
	There is a dog with no fleas. ★

As you can see from the above examples, forming the negation of a statement takes some care. For instance, notice that the negation of "Some cats have fleas" is "No cats have fleas", and not "Some cats do not have fleas". In forming negations, it pays to think about the meaning of the statement before placing the word "not" somewhere in the statement.

Example 2

The conjunction of the statements "I am a student of logic" and "I like to dispute" is

I am a student of logic and I like to dispute.

The disjunction of these statements is

I am a student of logic or I like to dispute.   ★

In order to translate a statement into symbolic form, we first assign variables to all of the simple statements that occur in the statement. Then we replace each of the five basic connectives by the symbols given in Table 1.

Example 3

Put the following sentences into symbolic form.
a)   I like studying and I like going to the library.
b)   Today is Friday or it is not Saturday.
c)   If I go to the movies, I will probably eat a lot of popcorn.
d)   I will go to the bank if and only if today is Friday.

Solutions
a)   If we let p be the statement "I like studying" and q be the statement "I like going to the library", then statement (a) has the symbolic form

$$p \wedge q$$

b)   If we let p be "Today is Friday" and q be "it is Saturday", then statement (b) has the symbolic form

$$p \vee (\sim q)$$

c)   If we let p be "I go to the movies" and q be "I will probably eat a lot of popcorn", then statement (c) has the symbolic form

$$p \rightarrow q$$

(Notice that the word "then" is implied in this sentence, even though it is not explicitly written.)

d)   If we let p be "I will go to the bank" and q be "today is Friday", then statement (d) has the symbolic form

$$p \leftrightarrow q \qquad \qquad \bigstar$$

Example 4

Consider the statement

I like logic, but not proofs.

This sentence has the same meaning as "I like logic *and* I don't like proofs". Hence, if we let p be "I like logic" and q be "I like proofs", we get the symbolic form

$$p \wedge (\sim q) \qquad \qquad \bigstar$$

Example 5

Let p be the simple statement "I study hard" and let q be the simple statement "I will get a good grade in this class". Then the symbolic statement

$$(p \rightarrow q) \wedge (\sim p \rightarrow \sim q)$$

translates into the English statement

If I study hard, then I will get a good grade in this class, and if I do not study hard, then I will not get a good grade in this class. $\bigstar$

Let us conclude this section with one last bit of terminology. The variables that we use in writing a statement in symbolic form are referred to as **statement variables**. We generally use the letters p, q, r, s and so on, to denote statement variables.

EXERCISES

Define or discuss each of the following terms.

a) Statement
b) Truth value
c) Connective
d) Simple statement
e) Compound statement
f) Conjunction
g) Disjunction
h) Negation
i) Conditional
j) Biconditional
k) Statement variable

1. Without looking in the text, write down each of the five basic connectives, along with its symbol and name.

In Exercises 2-11, determine whether the given sentence is a statement. If so, determine whether it is simple or compound.

2. No smoking.
3. All men are created equal.
4. Did you take out the garbage?
5. I love logic, but I hate algebra.
6. I will go if you want me to go.
7. I will get a new car if and only if I get a job.
8. The judge said he is guilty.
9. Stop talking and pay attention.
10. You must take the test today or tomorrow.
11. If today is Saturday, then I will go to the library.

In Exercises 12-18, find the negation of the given statement. Write your answer in grammatically correct English.

12. All horses can run.
13. There is a horse that can run.
14. There is a horse that cannot run.
15. There are no flowers in the garden.
16. $x - 4 \neq 3$
17. Some cats have kittens.
18. It is not true that today is Sunday.

In Exercises 19-23, let p be the statement "Romeo loves Juliet", and let q be "Juliet loves Romeo". Translate the given statement into symbolic form.

19. Romeo loves Juliet and Juliet loves Romeo.
20. Romeo loves Juliet or Juliet loves Romeo.
21. Juliet loves Romeo, but Romeo does not love Juliet.
22. If Romeo loves Juliet, then Juliet loves Romeo.
23. Romeo loves Juliet if and only if Juliet loves Romeo.

*In Exercises 24-27, let p be "I am a student," and let q be "I love math."
Translate the given sentence into symbolic form.*

24. I am not a student.
25. I am not a student nor do I love math.
26. It is not true that I am a student and I love math.
27. If I am a student, I love math.

*In Exercises 28-34, let p be the statement "help is coming" and let q be the
statement "the ship is sinking". Translate the given statement into a
grammatically correct English sentence.*

28. $p \wedge q$ 29. $q \vee \sim p$ 30. $q \rightarrow p$
31. $p \rightarrow \sim q$ 32. $q \leftrightarrow p$ 33. $\sim (\sim q)$
34. $q \rightarrow (p \vee \sim p)$

*In Exercises 35-40, let p be the statement "all birds are animals," let q be the
statement "some birds are animals," and let r be the statement "some animals
are birds." Translate the given statement into a grammatically correct English
sentence.*

35. $\sim p$ 36. $\sim q$ 37. $\sim r$
38. $\sim q \rightarrow \sim p$ 39. $p \wedge q \wedge \sim r$ 40. $p \wedge (q \vee r)$

*In Exercises 41-50, translate the statement into symbolic form. Indicate what
each statement variable means.*

41. If wishes were dollars, I'd be a millionaire.
42. Roses are red and violets are blue.
43. I will go to school if and only if today is Friday.
44. I love logic and I hate chemistry.
45. If you do not study, you cannot go to the movies.
46. You can't have dessert until you finish your vegetables.
47. To be or not to be.
48. Either you go to class or you stay home and study.
49. If you eat that chocolate sundae, then you will get sick, and fat too.
50. I'll be tired in the morning if I don't go to sleep now.

1.2   Truth Tables

Now that we know how to translate statements into symbolic form, we can address the issue of how to tell whether a given statement is true or false.

It is important, however, to make it clear that mathematical logic can never tell us whether a *simple* statement is true or false $-$ this must come from our knowledge of the given situation. For example, logic cannot tell us whether the statement "Today is Monday" is true or not. For this, we must look at a calendar.

On the other hand, mathematical logic *can* tell us whether a *compound* statement is true or false, if we already know the truth values of the simple statements that make it up. This is done by means of *truth tables*, which give us the precise meaning of each of the five basic connectives. Let us look at the truth tables of the connectives. (In these tables, we use T for "True" and F for "False.")

Negation

The truth table that describes the precise logical meaning of negation is

p	$\sim p$
T	F
F	T

This table certainly captures the spirit of the word *not*, in that a statement p is true if and only if $\sim p$ is false.

Conjunction

The truth table for conjunction is

p	q	$p \wedge q$
T	T	T
T	F	F
F	T	F
F	F	F

This captures the meaning of the word *and*, in that $p \wedge q$ is true when and only when *both p and q* are true.

Disjunction

The truth table for disjunction is

p	q	$p \vee q$
T	T	T
T	F	T
F	T	T
F	F	F

This captures the meaning of the word *or*, in that $p \vee q$ is true precisely when either *p or q*, *or both*, are true.

In everyday life, there are two common interpretations for the phrase "p or q". One is "p or q or both" and the other is "p or q but not both". The former interpretation of "or" is referred to as the **inclusive or**, and the latter interpretation is referred to as the **exclusive or**. Unfortunately, in everyday conversation, it is not always possible to determine which interpretation is intended. However, *in mathematical logic, the term disjunction always refers to the inclusive or*. Thus, for us, the statement "p or q" will always mean "p or q or both".

Conitional

The truth table for the conditional is

p	q	$p \rightarrow q$
T	T	T
T	F	F
F	T	T
F	F	T

The truth table for the conditional is not quite as "obvious" as the tables for the other connectives. Certainly, if p and q are true, one would expect the statement "if p then q" to also be true. Similarly, if p is true and q is false, then we certainly want the statement "if p then q" to be false, for we cannot be allowed to deduce a false q from a true p. However, the last two rows of the table are not so evident. There are many ways to justify this table, and we will give one such in the exercises, but for now let us simply say that this table has become the accepted meaning for the conditional, and that, as you will see, it works just fine. (It does have the advantage that it is easy to remember — all you have to do is remember that $p \rightarrow q$ is false when and *only* when p is true and q is false.)

Biconditional

The truth table for the biconditional is

p	q	$p \leftrightarrow q$
T	T	T
T	F	F
F	T	F
F	F	T

This table reflects the fact that the statement $p \leftrightarrow q$ should be true when, and only when, p and q have the *same* truth value.

Now that we have the truth tables for the five basic connectives, we can use them to obtain the truth table for any statement. This will enable us to determine precisely when any statement is true or false.

Example 1

The truth table for the statement $p \vee (\sim q)$ is

p	q	$\sim q$	$p \vee (\sim q)$
T	T	F	T
T	F	T	T
F	T	F	F
F	F	T	T

Notice that, in order to obtain the truth values for this statement, we first had to find the truth values of the "intermediate" statement $\sim q$. This was done in the third column of the table. It is typical for truth tables to have one or more intermediate columns.★

Example 2

The truth table for the statement $q \rightarrow (p \wedge \sim q)$ is

p	q	$\sim q$	$p \wedge \sim q$	$q \rightarrow (p \wedge \sim q)$
T	T	F	F	F
T	F	T	T	T
F	T	F	F	F
F	F	T	F	T

The fourth column of this table was obtained from the first and third columns using the truth table for the conjunction.★

Example 3

The truth table for the statement $(r \wedge q) \leftrightarrow (p \wedge q)$ is

p	q	r	$r \wedge q$	$p \wedge q$	$(r \wedge q) \leftrightarrow (p \wedge q)$
T	T	T	T	T	T
T	T	F	F	T	F
T	F	T	F	F	T
T	F	F	F	F	T
F	T	T	T	F	F
F	T	F	F	F	T
F	F	T	F	F	T
F	F	F	F	F	T

As you can see from these examples, finding the truth table of a statement is just a matter of repeatedly using the truth tables of the five basic connectives to build, in a step-by-step manner, the truth table of the given statement. *For this reason, we suggest that you learn the truth tables for the basic connectives extremely well.*

Notice that since the statements in Examples 1 and 2 involve two variables, the truth tables require 4 rows in order to express the 4 possibilities for the truth values of the variables. The statement in Example 3 involves 3 variables, and the truth table requires 8 rows to express all of the possibilities. (In constructing tables with 3 variables, it may help to observe that the first column consists of alternating groups of four T's and F's, the second column consists of alternating groups of two T's and F's, and the third column consists of alternating single T's and F's.)

In general, if a statement involves n distinct variables, then its truth table requires 2^n rows. This follows from the fact that there are 2 possibilities for the truth values of each of the variables, and so there are

$$\underbrace{2 \cdot 2 \cdots 2}_{n \text{ factors}} = 2^n$$

possibilities in total.

Example 4

To find the truth table for the statement

If oil is spilled at sea,
we must either clean it up or the planet will suffer.

we let

$p =$ oil is spilled at sea
$q =$ we must clean the oil up
$r =$ the planet will suffer

Then the symbolic form of our statement is

$$p \rightarrow (q \vee r)$$

whose truth table is

p	q	r	$q \vee r$	$p \rightarrow (q \vee r)$
T	T	T	T	T
T	T	F	T	T
T	F	T	T	T
T	F	F	F	F
F	T	T	T	T
F	T	F	T	T
F	F	T	T	T
F	F	F	F	T

From this, we see that the only time that this statement is false is when p is true, q is false, and r is false, that is, when oil is spilled at sea, we don't clean it up, and the planet doesn't suffer.★

Tautologies and Contradictions

Now let us consider the truth table for the statement $p \rightarrow (p \vee q)$,

p	q	$p \vee q$	$p \to (p \vee q)$
T	T	T	T
T	F	T	T
F	T	T	T
F	F	F	T

Notice that the last column of this table consists entirely of T's, and so the statement $p \to (p \vee q)$ is always true, regardless of the truth values of p and q. When this happens, we refer to the statement as a *tautology*. Let us make a formal definition of this important concept.

Definition

A statement that is always true, regardless of what truth values are assigned to the variables (or simple statements) that make it up, is called a **tautology**. We also say that such a statement is **logically true.**★

Consider now the truth table for $p \wedge (\sim p \wedge q)$,

p	q	$\sim p$	$\sim p \wedge q$	$p \wedge (\sim p \wedge q)$
T	T	F	F	F
T	F	F	F	F
F	T	T	T	F
F	F	T	F	F

In this case, the statement is always false. Such statements are called *contradictions*.

Definition

A statement that is always false, regardless of what truth values are assigned to the variables (or simple statements) that make it up, is called a **contradiction**. We also say that such a statement is **logically false.**★

It is important to understand the distinction between a statement that is *logically* true (or a tautology) and one that is merely true. For example, the statement

Today is Friday and it is raining

is true *provided* that both of the statements "Today is Friday" and "it is raining" are true. Thus, it is true on a rainy Friday, but false at all other times, and so it is *not* logically true. Notice that, in order to determine the truth value of this statement, we must examine its *meaning* at the time in question.

On the other hand, the statement

Today is Friday or today is not Friday

is always true, whether or not it is actually Friday. In other words, it is logically true. In this case, the statement is true, not because of its meaning, but because of its *structure*. Any statement of the form "*something* or not *something*" is true, regardless of the meaning of the "something".

Here are two more examples of tautologies, along with their symbolic forms

If all apples are green, then all apples are green. $(p \rightarrow p)$
If today is Sunday and you wash your car, then you wash your car.
$(p \wedge q) \rightarrow q$

Another Look at the Conditional

Let us conclude this section with another look at the conditional. According to the truth table for the conditional, if p is false, then the statement $p \rightarrow q$ is true. This can lead to some rather surprising consequences. For instance, since $2 + 2 = 5$ is false, the statement

If $2 + 2 = 5$, then the moon is made of green cheese

is true! As another example, the statement

Every person in your math class over 15 feet tall has four eyes

is also true, since this statement can be reworded in the form

If a person in your math class is over 15 feet tall,
then that person has four eyes

and this has the symbolic form $p \rightarrow q$, where p is false, since no person in your math class is over 15 feet tall.

EXERCISES

Define or discuss each of the following terms.

a) Negation   b) Conjunction
c) Disjunction   d) Conditional
e) Biconditional   f) Tautology
g) Contradiction

In Exercises 1-21 , find the truth table for the given statement.

1. $\sim(\sim p)$
2. $p \rightarrow \sim p$
3. $\sim(\sim p) \rightarrow \sim(\sim p \rightarrow p)$
4. $p \wedge \sim q$
5. $p \wedge (p \vee q)$
6. $p \wedge (\sim p \vee \sim q)$
7. $p \rightarrow (p \vee \sim p)$
8. $p \rightarrow (p \vee \sim q)$
9. $(p \wedge q) \rightarrow (p \vee q)$
10. $(p \rightarrow \sim p) \rightarrow (q \rightarrow \sim q)$

11. $(p \leftrightarrow q) \wedge (p \leftrightarrow r)$
12. $(p \wedge q) \rightarrow r$
13. $(p \rightarrow r) \rightarrow (q \rightarrow r)$
14. $(p \wedge q) \vee r$
15. $(\sim p \wedge \sim q) \vee r$
16. $[(p \rightarrow q) \wedge (q \rightarrow p)] \leftrightarrow (p \leftrightarrow q)$
17. $(p \rightarrow \sim q) \wedge (p \rightarrow \sim r)$
18. $[p \rightarrow (q \rightarrow r)] \rightarrow [(p \rightarrow q) \rightarrow (p \rightarrow r)]$
19. $(p \wedge q) \vee (r \wedge s)$
20. $(p \rightarrow q) \wedge (r \rightarrow s)$
21. $p \wedge (\sim r) \wedge s \wedge (\sim q)$
22. Show that the statement $\sim p \rightarrow (p \rightarrow q)$ is a tautology.
23. Show that the statement $(\sim p \rightarrow p) \rightarrow p$ is a tautology.
24. Show that the statement $p \rightarrow (q \rightarrow p)$ is a tautology.
25. Show that the statement $(p \wedge q) \leftrightarrow (q \wedge p)$ is a tautology.
26. Show that the statement $(\sim p \rightarrow \sim q) \rightarrow (q \rightarrow p)$ is a tautology.
27. a)   Show that the statement $\sim (p \wedge q) \leftrightarrow (\sim p \vee \sim q)$ is a tautology.
    b)   Show that the statement $\sim (p \vee q) \leftrightarrow (\sim p \wedge \sim q)$ is a tautology.
28. Show that the statement $(p \wedge q) \wedge \sim (p \vee q)$ is a contradiction.
29. Show that the statements $p \rightarrow q$ and $\sim p \vee q$ have the same truth table. What do you make of this?
30. Show that the statements $p \leftrightarrow q$ and $(p \rightarrow q) \wedge (q \rightarrow p)$ have the same truth table. What do you make of this?
31. Find the truth table for the exclusive or.
32. Is the statement $p \rightarrow \sim p$ a contradiction? Do you think it should be?
33. Suppose that A and B are compound statements. If A is a contradiction, what can you say about the statement $A \rightarrow B$?
34. Let A, B and C be statements. If $A \rightarrow B$ and $B \rightarrow C$ are tautologies, can you show that $A \rightarrow C$ is a tautology?
35. Now let us discuss one reason why the truth table for the conditional is defined the way it is. As we discussed in the text, the first two rows of the truth table are quite reasonable. With this in mind, we are left with 4 possibilities for the last two rows, as shown in the following table.

Possibilities for $p \rightarrow q$					
p	q	(1)	(2)	(3)	(4)
T	T	T	T	T	T
T	F	F	F	F	F
F	T	T	T	F	F
F	F	T	F	T	F

Note that the conditional is actually defined using column (1) of this table. Now, it seems intuitively clear that the statement $p \rightarrow (p \vee q)$ should always be true; that is, it should be a tautology. But, we ask you to show that this statement is a tautology if the conditional is defined by column (1), but not if we were to use any of the other three columns for the definition of the conditional.

1.3 Logical Equivalence

At this point, we have developed the necessary tools to decide what it means for two statements to have the same logical meaning, that is, to be *logically equivalent.*

Definition

We say that two statements A and B are **logically equivalent** if they have the same truth value, for all possible values of the statement variables. When A and B are logically equivalent, we denote this by writing $A \equiv B$.★

Since the truth table of a statement gives us the truth values of the statement for all possible values of the variables, we can tell whether or not two statements are logically equivalent simply by comparing their truth tables!

For example, consider the statements $p \rightarrow q$ and $\sim p \vee q$. Their truth tables are

p	q	$p \rightarrow q$	p	q	$\sim p$	$\sim p \vee q$
T	T	T	T	T	F	T
T	F	F	T	F	F	F
F	T	T	F	T	T	T
F	F	T	F	F	T	T

Since these tables are the same (except for intermediate columns), the statements $p \rightarrow q$ and $\sim p \vee q$ are logically equivalent, that is, they have the same logical meaning. Thus, we may write

$$p \rightarrow q \equiv \sim p \vee q$$

DeMorgan's Laws

Recalling our discussion in the Introduction, we can use the concept of logical equivalence to help understand why the statement

$$\text{It is not true that I have a cat and I have a dog.} \tag{1}$$

has the same logical meaning as the statement

$$\text{I do not have a cat } or \text{ I do not have a dog.} \tag{2}$$

and not

$$\text{I do not have a cat } and \text{ I do not have a dog.} \tag{3}$$

as many people think.

First, we translate each of these statements into symbolic form

$$\sim (p \wedge q) \tag{1'}$$

$$(\sim p) \vee (\sim q) \tag{2'}$$

$$(\sim p) \wedge (\sim q) \tag{3'}$$

Now, we wish to determine which of (2′) or (3′) is logically equivalent to (1′), and so we construct the truth tables.

Truth Table for $(1')$			
p	q	$p \wedge q$	$\sim (p \wedge q)$
T	T	T	F
T	F	F	T
F	T	F	T
F	F	F	T

Truth Table for $(2')$				
p	q	$\sim p$	$\sim q$	$(\sim p) \vee (\sim q)$
T	T	F	F	F
T	F	F	T	T
F	T	T	F	T
F	F	T	T	T

Truth Table for $(3')$				
p	q	$\sim p$	$\sim q$	$(\sim p) \wedge (\sim q)$
T	T	F	F	F
T	F	F	T	F
F	T	T	F	F
F	F	T	T	T

Since the first two truth tables are the same, we see that statements (1′) and (2′) are logically equivalent. Also, since the first and third truth tables are *not* the same, we see that statements (1′) and (3′) are not logically equivalent. In other words, statements (1) and (2) have the same meaning, but statements (1) and (3) do not.

The fact that (1′) and (2′) are logically equivalent is an example of *DeMorgan's Laws*.

Theorem 1

(DeMorgan's Laws) For any two statements A and B, we have
1) $\sim (A \wedge B) \equiv (\sim A) \vee (\sim B)$
2) $\sim (A \vee B) \equiv (\sim A) \wedge (\sim B)$ ★

Let us consider some additional examples of DeMorgan's laws.

Example 1

The symbolic form of the statement

> I will study alone or I will hire a tutor

is $p \vee q$, where p is "I will study alone" and q is "I will hire a tutor". Hence, its negation is $\sim (p \vee q)$, which is logically equivalent to $(\sim p) \wedge (\sim q)$, which in turn translates into

> I will not study alone *and* I will not hire a tutor.   ★

Example 2

The symbolic form of the statement

> Most students are interested in logic and are not interested in chemistry

is $p \wedge \sim q$, where p is "Most students are interested in logic" and q is "Most students are interested in chemistry". Hence, its negation is

$$\sim (p \wedge \sim q)$$

which, by DeMorgan's Laws, is logically equivalent to

$$\sim p \vee \sim (\sim q)$$

But since $\sim (\sim q) \equiv q$, we can simplify this to

$$\sim p \vee q$$

This translates into

> Most students are not interested in logic *or* they are interested in chemistry. ★

The Converse of a Statement

Another issue that is often misunderstood by those who have not studied logic is the relationship between the statements of the form $A \rightarrow B$ and $B \rightarrow A$. Let us begin with a definition.

Definition

Let A and B be any statements, simple or compound. Then
1) The **converse** of the statement $A \rightarrow B$ is the statement $B \rightarrow A$.
2) The **inverse** of the statement $A \rightarrow B$ is the statement $\sim A \rightarrow \sim B$.
3) The **contrapositive** of the statement $A \rightarrow B$ is the statement $\sim B \rightarrow \sim A$.

For easy reference, let us put these terms into a table

A statement:	$A \to B$
Its **converse**:	$B \to A$
Its **inverse**:	$\sim A \to \sim B$
Its **contrapositive**:	$\sim B \to \sim A$

Here is an example.

Example 3

Consider the statement

$$\text{If today is Tuesday, then I have math class.} \qquad (4)$$

This statement has the symbolic form $p \to q$, where p is "Today is Tuesday" and q is "I have math class".

The converse of $p \to q$ is $q \to p$, and so the converse of (4) is

$$\text{If I have math class, then today is Tuesday.}$$

The inverse of $p \to q$ is $\sim p \to \sim q$, and so the inverse of (4) is

$$\text{If today is not Tuesday, then I do not have math class.}$$

The contrapositive of $p \to q$ is $\sim q \to \sim p$, and so the contrapositive of (4) is

$$\text{If I do not have math class, then today is not Tuesday.} \qquad \bigstar$$

To understand the relationship between a statement and its converse, inverse and contrapositive, we construct a truth table that has a column for each statement

				Statement	Converse	Inverse	Contrapositive
p	q	$\sim p$	$\sim q$	$p \to q$	$q \to p$	$\sim p \to \sim q$	$\sim q \to \sim p$
T	T	F	F	T	T	T	T
T	F	F	T	F	T	T	F
F	T	T	F	T	F	F	T
F	F	T	T	T	T	T	T

This table tells us that a statement and its *contrapositive* are logically equivalent, but that a statement and its converse are not logically equivalent. Thus, we have the following theorem.

Theorem 2

1) A statement and its contrapositive are logically equivalent.
2) The converse and the inverse are logically equivalent.
3) No other logical equivalences exist between a statement, its converse, inverse and contrapositive. In particular, a statement is *not* equivalent to its converse. $\bigstar$

Example 4

Consider the statement

$$\text{If it is raining, then I will not go to the park.} \tag{5}$$

This has the symbolic form $p \to \, \sim q$, where p is "It is raining" and q is "I will go to the park".

The converse of $p \to \, \sim q$ is $\sim q \to p$, and so the converse of (5) is

$$\text{If I do not go to the park, then it is raining.} \tag{6}$$

According to Theorem 2, and contrary to what many people believe, this statement does *not* have the same meaning as (5).

The inverse of $p \to \, \sim q$ is $\sim p \to \, \sim (\sim q)$. But $\sim (\sim q)$ is logically equivalent to q, and so we can simplify this to $\sim p \to q$. Hence, the inverse of (5) is

$$\text{If it is not raining, then I will go to the park.}$$

According to Theorem 2, this has the same meaning as statement (6), but not as the original statement (5).

The contrapositive of $p \to \, \sim q$ is $\sim (\sim q) \to \, \sim p$, and as before, we can replace $\sim (\sim q)$ by q, to get $q \to \, \sim p$. Hence, the contrapositive of (5) is

$$\text{If I go to the park, then it is not raining.}$$

According to Theorem 2, this statement has the same meaning as the original statement (5). ★

EXERCISES

Define or discuss each of the following terms.

a) Logical equivalence   b) Converse
c) Inverse   d) Contrapositive
e) DeMorgan's Laws

In Exercises 1-8, use truth tables to verify that the given statements are logically equivalent.

1. $p \equiv \, \sim (\sim p)$
2. $p \to q \equiv (p \wedge \sim q) \to \, \sim p$
3. $p \to q \equiv (p \wedge \sim q) \to q$
4. $p \leftrightarrow q \equiv (p \to q) \wedge (q \to p)$
5. $p \wedge q \equiv q \wedge p$
6. $p \equiv p \wedge p$
7. $p \equiv p \vee (p \wedge q)$
8. $p \wedge (q \vee r) \equiv (p \wedge q) \vee (p \wedge r)$

In Exercises 9-16, write the negation of each statement in grammatically correct English.

9. a) Roses are red and violets are blue.

    b)   Roses are red or violets are blue.
10. a)   Elephants are large but flies are small.
    b)   Elephants are large, or flies are small.
11. a)   Elephants are large and flies are not.
    b)   Elephants are large, or flies are not.
12. a)   Today is not Friday and it is raining.
    b)   Today is not Friday or it is raining.
13. a)   Today is Friday and Saturday.
    b)   Today is Friday or Saturday.
14. a)   It is not raining and it is not snowing.
    b)   It is not raining or it is not snowing.
15. a)   It is neither Friday nor Saturday.
    b)   It is not Friday or not Saturday.
16. a)   It is not true that birds can fly and fish can swim.
    b)   It is not true that birds can fly or fish can swim.

In Exercises 17-28, find the converse, inverse and contrapositive of the given statement. If the statement is in English, then your answers should be in grammatically correct English.

17. $\sim p \to q$

18. $\sim p \to \sim q$

19. $(p \wedge q) \to r$

20. $(p \vee q) \to \sim p$

21. $(p \to q) \to (r \to s)$

22. If war continues, then more people will die.

23. If Abraham Lincoln were alive today, he would be elected president.

24. If today is a holiday, then I don't have to go to school.

25. If today is not a holiday, then I have to go to school.

26. If x is even, then so is 3x.

27. If today is Monday and it is raining, then I am not going to school.

28. If today is Monday and it is not raining, then I am going to school.

29. Let A and B be statements. We say that A **logically implies** B if B is true whenever A is true.

    a)   What does this mean in terms of the truth tables for A and B?

    b)   Show that $p \wedge q$ logically implies $p \vee q$, but that these statements are not logically equivalent.

Chapter 2
Applications of Logic

2.1 Valid Arguments

Now we are ready to discuss the issue of how to tell whether a given argument is valid or not. Loosely speaking, an *argument* is just a sequence of statements, called *premises*, or *hypotheses*, followed by another statement, called the *conclusion*. The idea, of course, is that the conclusion is supposed to follow logically from the hypotheses, and this is what we mean by a *valid* argument. Let us give a more precise definition of these terms.

Definition

An **argument** is a sequence of statements, written in the form

$$A_1, A_2, \ldots, A_n \therefore A$$

or

$$
\begin{array}{c}
A_1 \\
A_2 \\
\vdots \\
A_n \\
\hline
A
\end{array}
$$

The statements $A_1, A_2, \ldots, A_n$ are called the **premises**, or **hypotheses**, of the argument, and the statement A is called the **conclusion** of the argument. The symbols $\therefore$ and the horizontal line —— both stand for the word "therefore." ★

For example, the statements

If today is a holiday, then I can go to the beach.
Today is a holiday.
Therefore, I can go to the beach.

constitute an argument. The first two statements are the hypotheses of the argument, and the third statement is the conclusion.

Notice that, even though an argument is not of much use if the hypotheses are not all true, according to the definition, we do not *require* that they all be true. Let us now define what it means for an argument to be valid.

Definition

An argument $A_1, A_2, \ldots, A_n \therefore A$ is **valid** if *whenever* the premises $A_1, A_2, \ldots, A_n$ are true, so is the conclusion A. If an argument is not valid, we say that it is **invalid**. ★

It is very important to understand this definition completely. According to the definition, to determine whether an argument is valid, we need *only* consider the cases where all of the premises are true. There will generally be cases where one or more of the premises are false. However, in such cases, we do not care whether the conclusion is true or false, for these cases have no bearing on the validity of the argument.

Determining whether an argument is valid or not can usually be done simply by constructing a giant truth table giving the truth values for each of the premises, as well as the conclusion. Let us illustrate with some examples.

Example 1

Consider the following argument

If today is a holiday, then I can go to the beach.
Today is a holiday.
Therefore, I can go to the beach.

The first step in determining validity is to express this argument in symbolic form. So we let p be "today is a holiday", and q be "I can go to the beach". Then the symbolic form of this argument is

$$p \rightarrow q$$
$$p$$
$$\overline{}$$
$$q$$

Next we construct a truth table that contains truth values for the two premises p and $p \rightarrow q$, as well as the conclusion q.

premise	conclusion	premise
p	q	$p \rightarrow q$
T	T	T
T	F	F
F	T	T
F	F	T

Now, row 1 of this table is the only row in which both premises are true, so this is the only row we need to be concerned about in determining the validity of the argument. Since in this row, the conclusion is also true, we can say that whenever all of the premises are true, so is the conclusion. Hence, the argument is valid. ★

The argument in the previous example has the general form

$$A \rightarrow B$$
$$A$$
$$\overline{}$$
$$B$$

where A and B are any statements (simple or compound.) This type of argument is one of the most commonly used arguments in debating, and was well-known even to Aristotle, who referred to it by the name **modus ponens**. (*Modus* is from the Latin meaning *in the manner of*, and *ponens* is from the Latin verb *pono*, meaning *to assert*. Thus, *modus ponens* means *in the manner of asserting*.) Let us have another example of modus ponens.

Example 2

Consider the argument

> If I study, I won't flunk the next test.
> I study.
> Therefore, I won't flunk the next test.

Here we let p be "I study" and q be "I flunk". Then the symbolic form of this argument is

$$p \rightarrow \sim q$$
$$p$$
$$\overline{}$$
$$\sim q$$

Since this is an example of modus ponens (where $A = p$ and $B = \sim q$), we know it is a valid argument. ★

Example 3

Consider the argument

> If today is a holiday, then I can go to the beach.
> I can go to the beach.
> Therefore, today is a holiday.

Letting p be "today is a holiday", and q be "I can go to the beach", we get the symbolic form

$$p \rightarrow q$$
$$q$$
$$\overline{}$$
$$p$$

This argument may look similar to modus ponens, but it is not the same, since the second premise and the conclusion are reversed. The truth table for this argument is

conclusion	premise	premise
p	q	$p \rightarrow q$
T	T	T
T	F	F
F	T	T
F	F	T

In this case, rows 1 and 3 have the property that both premises are true. However, since in row 3 the conclusion is false, we see that it is possible for all of the premises to be true and yet the conclusion can be false. Hence, this argument is *invalid.* ★

Example 4

Consider the argument

> If today is a holiday, then I can go to the beach.
> If I can go to the beach, then I can take a surfing lesson.
> Therefore, if today is a holiday, I can take a surfing lesson.

Letting p be "today is a holiday", q be "I can go to the beach", and r be "I can take a surfing lesson", we get the symbolic form

$$p \rightarrow q$$
$$q \rightarrow r$$
$$\overline{}$$
$$p \rightarrow r$$

To test the validity of this argument, we construct the following truth table

p	q	r	premise $p \to q$	premise $q \to r$	conclusion $p \to r$
T	T	T	T	T	$T \leftarrow$
T	T	F	T	F	F
T	F	T	F	T	T
T	F	F	F	T	F
F	T	T	T	T	$T \leftarrow$
F	T	F	T	F	T
F	F	T	T	T	$T \leftarrow$
F	F	F	T	T	$T \leftarrow$

The rows marked with an arrow are the rows for which all of the premises are true. Since for each of these rows, the conclusion is also true, we see that this argument is valid. ★

Example 5

Consider the argument

$$(p \wedge q) \to r$$
$$p$$
$$\overline{}$$
$$r$$

The appropriate truth table in this case is

premise		conclusion		premise
p	q	r	$p \wedge q$	$(p \wedge q) \to r$
T	T	T	T	$T \leftarrow$
T	T	F	T	F
T	F	T	F	$T \leftarrow$
T	F	F	F	$T \leftarrow$
F	T	T	F	T
F	T	F	F	T
F	F	T	F	T
F	F	F	F	T

The marked rows are the only ones for which all premises are true. However, in row 4 we observe that the conclusion is false, and so this argument is invalid. ★

Sorites

The argument in Example 4 has the general form

$$A \to B$$
$$B \to C$$
$$\overline{A \to C}$$

where A, B and C are statements. More generally, arguments of the form

$$(1) \quad \begin{array}{l} A_1 \to A_2 \\ A_2 \to A_3 \\ \quad\vdots \\ A_{n-1} \to A_n \\ \hline A_1 \to A_n \end{array}$$

where $A_1, \ldots, A_n$ are statements (simple or
(pronounced with a long
"chaining" effect in the
argument.)

Example 6

Consider the argument

A student who studies w
John is very smart.
Even smart people must
Therefore, John will not f

This argument is a bit more c   than the previous ones, partly due to
the fact that the language is a bit more convoluted. In such a case, it may help to
first "condense" the meaning of the argument, as follows

Study implies not flunk
John implies smart
Smart implies study
Therefore, John implies not flunk

Now if we choose our statement variables as follows

p be "study"
q be "flunk"
r be "smart"
s be "John"

then this argument has the symbolic form

$$p \rightarrow \sim q$$
$$s \rightarrow r$$
$$r \rightarrow p$$
$$\overline{}$$
$$s \rightarrow \sim q$$

Rearranging the order of the hypotheses certainly will not affect the validity of the argument, and gives

$$s \rightarrow r$$
$$r \rightarrow p$$
$$p \rightarrow \sim q$$
$$\overline{}$$
$$s \rightarrow \sim q$$

But this is a sorites (note the chaining in the hypotheses) and so we know that it is valid.★

Lewis Carroll, the author of *Alice in Wonderland*, was a great fan of symbolic logic. In fact, he wrote a charming little book entitled *Symbolic Logic* (published by Dover Publications), in which he gave 60 humorous examples of sorites.

Lewis Carroll
(1832-1898)

Before looking at an example, we should observe that replacing a hypothesis in a given argument by a logically equivalent statement does not effect the validity of the argument. This can be a very useful tool in determining the validity of an argument.

Example 7

Consider the following argument from Lewis Carroll's book

No ducks waltz.
No officers ever decline to waltz.
All my poultry are ducks.
Therefore, my poultry are not officers.

To obtain the symbolic form of this argument, we let

p be "ducks"
q be "willing to waltz"
r be "officers"
s be "my poultry"

Then Lewis Carroll's argument has the symbolic form

$$p \to \sim q$$
$$r \to q$$
$$s \to p$$
$$\overline{}$$
$$s \to \sim r$$

Rearranging the order of the premises gives

$$s \to p$$
$$p \to \sim q$$
$$r \to q$$
$$\overline{}$$
$$s \to \sim r$$

(We do this to try to get the chaining effect that is characteristic of a sorites.) Since a statement is logically equivalent to its contrapositive, we may replace any statement by its contrapositive, and not affect the validity of the argument. Thus, since the contrapositive of $r \to q$ is $\sim q \to \sim r$, we get the equivalent argument

$$s \to p$$
$$p \to \sim q$$
$$\sim q \to \sim r$$
$$\overline{}$$
$$s \to \sim r$$

Now we see the "chaining" effect that is present in a sorites. In particular, this argument has the form (1) with $A_1 = s$, $A_2 = p$, $A_3 = \sim q$ and $A_4 = \sim r$. Hence, it is valid. ★

Actually, in Lewis Carroll's book, the sorities are given without conclusions, and you are asked to fill in the conclusion to make a valid argument. Let us try one in this form.

Example 8

Consider the following premises to an argument

> All my sons are slim.
> No child of mine is healthy who takes no exercise.
> All gluttons, who are children of mine, are fat.
> No daughter of mine takes any exercise.

In order to determine the conclusion to this argument, we first assign statement variables as follows. Let

> p be "son"
> q be "is slim"
> r be "is healthy"
> s be "takes exercise"
> t be "glutton"

Notice that, in this argument, Lewis Carroll is talking about his children, and so "daughter" is the same as "not son", that is, "daughter" is represented by $\sim p$. Also, "fat" is represented by $\sim q$.

The symbolic form of our (partial) argument is

$$p \rightarrow q$$
$$\sim s \rightarrow \sim r$$
$$t \rightarrow \sim q$$
$$\sim p \rightarrow \sim s$$

Noticing that t appears only once, and recalling that we will be able to replace any of these statements by their contrapositives, we reorganize as follows

$$t \rightarrow \sim q$$
$$p \rightarrow q$$
$$\sim p \rightarrow \sim s$$
$$\sim s \rightarrow \sim r$$

Replacing the second statement by its contrapositive gives

$$t \rightarrow \sim q$$
$$\sim q \rightarrow \sim p$$
$$\sim p \rightarrow \sim s$$
$$\sim s \rightarrow \sim r$$

This is the beginning of a sorites, and we see that the conclusion is $t \rightarrow \sim r$, that is

All my gluttonous children are unhealthy!   ★

EXERCISES

Define or discuss the following terms.

a)   Argument   b)   Premise
c)   Hypothesis   d)   Valid argument
e)   Invalid argument

In Exercises 1-15, determine whether or not the given argument is valid.

1. $p \rightarrow q$
    $\sim q$
    ______
    $\sim p$

2. $p \rightarrow q$
    $\sim q$
    ______
    p

3. $p \vee q$
    $\sim p$
    ______
    q

4. $p \vee q$
    $q \rightarrow p$
    ______
    p

5. p
    $q \rightarrow \sim p$
    ______
    $\sim q$

6. $p \rightarrow q$
    $p \rightarrow r$
    ______
    $q \rightarrow r$

7. $p \rightarrow q$
    $q \wedge r$
    ______
    $r \rightarrow \sim q$

8. $p \rightarrow q$
    $\sim p \vee q$
    ______
    $q \rightarrow p$

9. $q \rightarrow p$
    $\sim q \leftrightarrow p$
    ______
    p

10. $(p \vee q) \rightarrow r$
     p
     ______
     r

11. $\sim r$
     $\sim q \rightarrow r$
     $\sim p \rightarrow r$
     ______
     $p \wedge q$

12. $(p \rightarrow q) \rightarrow r$
     q
     ______
     $p \rightarrow r$

13. $p \rightarrow (q \rightarrow r)$
     p
     ______
     $q \rightarrow r$

14. $p \rightarrow (q \rightarrow r)$
     q
     ______
     $p \rightarrow r$

15. $p \leftrightarrow q$
     $q \leftrightarrow r$
     ______
     $p \leftrightarrow r$

In Exercises 16-26, translate the argument into symbolic form and determine whether or not it is valid.

16. It is raining or it is snowing.
     It is not snowing.
     Therefore, it is raining.
17. It is raining or it is not snowing.
     It is not raining.
     Therefore, it is snowing.
18. It is raining or it is not snowing.
     It is snowing.
     Therefore, it is not raining.
19. I have math class if and only if today is Friday.
     I get the car if and only if today is Friday.
     Therefore, I get the car if and only if I have math class.
20. If I eat a lot of chocolate, I will gain weight.

If I gain weight, I will need to buy new clothes.
I need to buy new clothes.
Therefore, I have gained weight.

21. If we have a *glasnost*, then democracy is safe.
     If Gorbachev stays in office, we will have *glasnost*.
     Therefore, democracy is safe.

22. If Mr. Smith bought a house, then either he sold his car or he borrowed money from a bank.
     Mr. Smith has not borrowed money from a bank.
     Therefore, if Mr. Smith has not sold his car, he has not bought a house.

23. If it snows today, then the newspaper is right or the radio is wrong.
     It snowed today and the newspaper is wrong.
     Therefore, the radio is wrong.

24. (Lewis Carroll)
     Babies are illogical.
     Nobody is despised who can manage a crocodile.
     Illogical persons are despised.
     Therefore, babies cannot manage crocodiles.

25. (Lewis Carroll)
     No birds, except ostriches, are 9 feet high.
     There are no birds in this aviary that belong to anyone but me.
     No ostrich lives on mince-pie.
     I have no birds less than 9 feet high.
     Therefore, no bird in this aviary lives on mince-pie.

26. (Lewis Carroll)
     No potatoes of mine, that are new, have been boiled.
     All my potatoes in this dish are fit to eat.
     No unboiled potatoes of mine are fit to eat.
     Therefore, all my potatoes in this dish are old ones.

In Exercises 27 − 30, determine the conclusion to Lewis Carroll's argument.

27. Every one who is sane can do logic.
     No lunatics are fit to serve on a jury.
     None of your sons can do logic.
     Therefore,...?

28. No one takes in the *Times*, unless he is well-educated.
     No hedge-hogs can read.
     Those who cannot read are not well-educated.
     Therefore,...?

29. All puddings are nice.
     This dish is a pudding.
     No nice things are wholesome.
     Therefore,...?

30. All unripe fruit is unwholesome.
     All these apples are wholesome.
     No fruit, grown in the shade, is ripe.
     Therefore,...?

2.2 Logic Circuits

We mentioned in the Introduction that logic can be used to design circuits, such as those used in computers. The type of circuit we are referring to is known as an **AND-OR circuit**, and consists of three types of components, or *gates*. The **AND gate** has the form

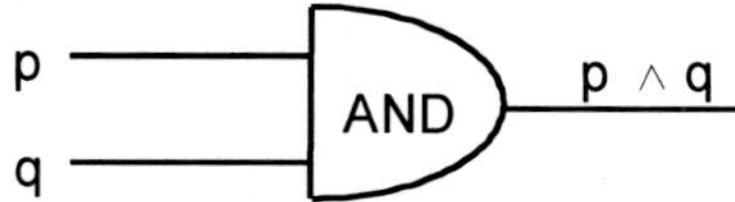

The wires leading into the AND gate are called the *inputs*, and the wire leading from the gate is called the *output* of the gate. All wires in a circuit carry either a high voltage, which we denote by T, or a low voltage, which we denote by F. As you can see from the label on the output wire, the output of an AND gate is the conjunction $p \wedge q$ of the inputs.

The **OR gate** is similar to the AND gate, but its output is the disjunction of the inputs. OR gates are drawn as follows

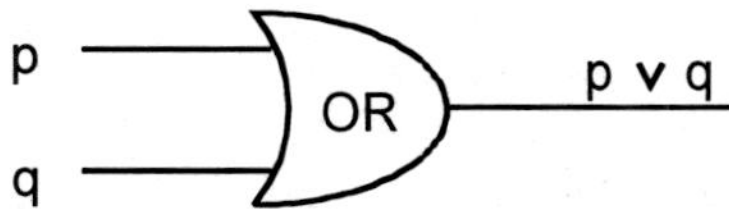

Our circuits also have a third type of gate, called a **NOT gate**, or **inverter**, drawn as follows

As the name implies, inverters change high voltage (T) to low voltage (F), and vice-versa. Notice also that we use a prime ($'$) to denote negation, as is customary for circuit design. (The prime takes up less space than the symbol $\sim$.)

In constructing AND-OR circuits, we will allow AND and OR gates to have more than two inputs. For example, consider the circuit in Figure 1.

Figure 1

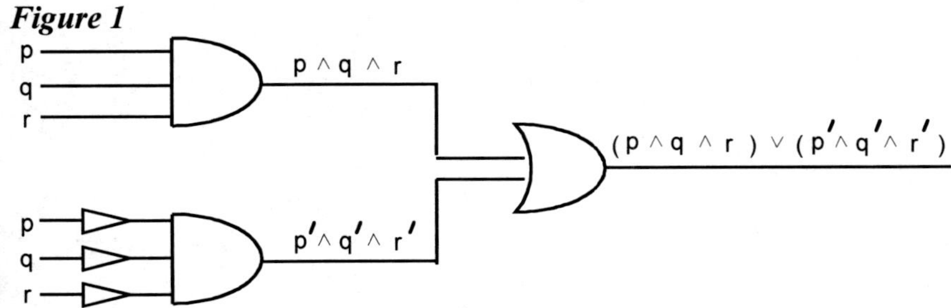

Notice that different input wires may be labeled with the same variable. Such wires always have the same input. For example, all input wires labeled p have the same input.

Notice also that we have labeled various wires on the circuit in Figure 1. The label on the output wire of the circuit (the last wire in the circuit) is the most important, and we refer to it as the **statement associated with the circuit**. Thus, the statement associated with the circuit in Figure 1 is

$$(p \wedge q \wedge r) \vee (p' \wedge q' \wedge r')$$

The truth table of this statement is called the **truth table of the circuit**

p	q	r	$p \wedge q \wedge r$	$p' \wedge q' \wedge r'$	$(p \wedge q \wedge r) \vee (p' \wedge q' \wedge r')$
T	T	T	T	F	T
T	T	F	F	F	F
T	F	T	F	F	F
T	F	F	F	F	F
F	T	T	F	F	F
F	T	F	F	F	F
F	F	T	F	F	F
F	F	F	F	T	T

The truth table of a circuit describes the "behavior" of the circuit. For instance, looking at the table, we see that the output of the circuit in Figure 1 is high voltage (T) if and only if either all inputs are high voltage $(p = T, q = T, r = T)$ or all inputs are low voltage $(p = F, q = F, r = F)$. In other words, this circuit outputs a high voltage precisely when the inputs are all the same. In the world of circuit design, this circuit is called a **comparator**, since it compares the inputs, and tells us whether or not they are the same!

Generally speaking, the goal in circuit design is to construct a circuit to perform a certain function, such as comparing the inputs. The first step in this process is to determine the truth table that describes the given behavior, and then construct the circuit from the table. Let us illustrate with two examples.

Example 1

Suppose we wish to construct a circuit that has three inputs p, q and r, and outputs a high voltage if and only if $p = q$ (r can be anything.) Thus, the truth table will have three variables p, q and r, and its last column will be T if and only if $p = q = T$, or $p = q = F$. Here is the table.

p	q	r	output
T	T	T	T
T	T	F	T
T	F	T	F
T	F	F	F
F	T	T	F
F	T	F	F
F	F	T	T
F	F	F	T

In order to design a circuit with this table as its truth table, we proceed as follows. For each row of the table with output T, we require one AND gate. There is one input to each AND gate for each variable, but those inputs corresponding to an F in the corresponding row of the table are inverted. To clarify this point, observe that rows 1, 2, 7 and 8 of the above table have output T. Hence, we require the following AND gates

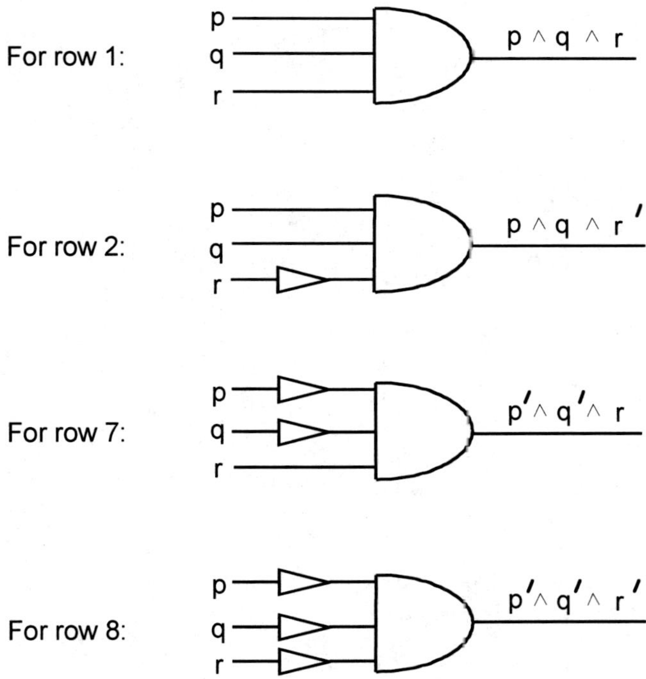

Looking at the AND gate for row 7, for instance, since the truth value of p and q in row 7 is F, we place an inverter on the input wires labeled p and q.

To complete the circuit, we run the outputs of the AND gates into an OR gate, as shown in Figure 2. This circuit outputs a high voltage if and only if $p = q$. ★

Figure 2

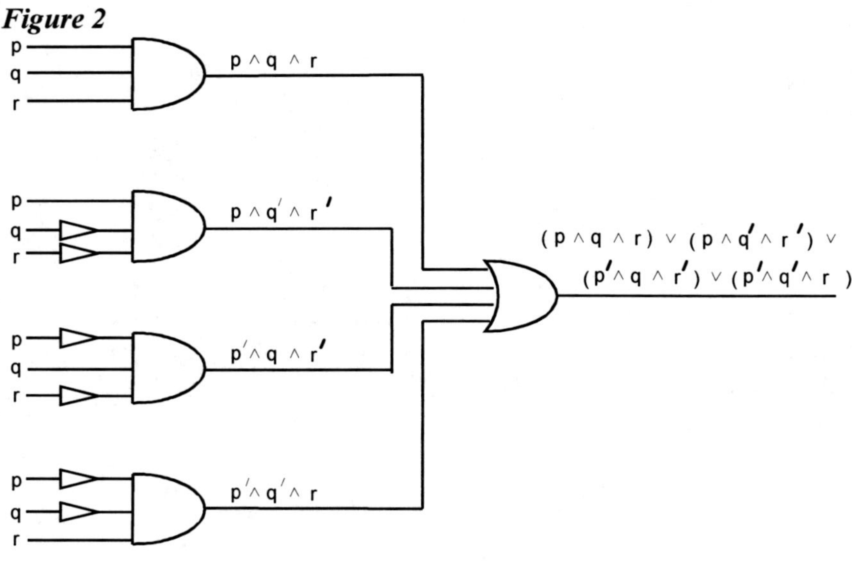

Example 2

It is common in rooms that have more than one entrance for there to be several light switches controlling a single overhead light. Somehow, it seems that each switch knows whether the light is on or off, because when you flip one of the switches, it changes the status of the light. In fact, sometimes the light is on when a given switch is up, and sometimes the light is off. Have you ever wondered how this is possible?

To answer this question, let us design a circuit with three switches that has precisely this behavior. To describe the circuit, we require three variables p, q and r to denote the status of the switches, and one variable L to denote the status of the light. A switch variable is true when the switch is on, and false when the switch is off. For instance, $p = T$, $q = T$ and $r = F$ means that switches p and q are on, and switch r is off. Further, the variable L is true when the light is on, and false when the light is off.

Now we can describe the behavior of the circuit in words. Whenever *exactly* one of the switches is changed, the light changes status. In terms of the variables, whenever *exactly* one of the variables p, q or r changes value, so does the variable L. Assuming that when all the switches are on ($p = T$, $q = T$, $r = T$), the light is on ($L = T$), we get the following table.

p	q	r	L
T	T	T	T
T	T	F	F
T	F	T	F
T	F	F	T
F	T	T	F
F	T	F	T
F	F	T	T
F	F	F	F

Let us explain in more detail how this table was constructed. The first row corresponds to our assumption that when all switches are on, so is the light. Since the second row is obtained from the first by changing exactly one switch (switch r), the light must change, and so for this row we must have $L = F$. Now, the third row also comes from the first row by changing a single switch (switch q), and so the light must again be off. Notice, however, that we must also check that rows 3 and 2 are consistent. The point is that row 3 can also be obtained from row 2 by making exactly *two* switch changes (switches q and r). Since two switch changes will leave the light unchanged, the status of the light must be the same for rows 2 and 3, which indeed it is. Had the light been different for rows 2 and 3, we would be forced to conclude that no circuit could be designed to do what we wish! Continuing in this way, that is, constructing each row based on row 1, and then checking it for consistency against all previous rows, we get the table shown above.

Once we have the truth table with the desired behavior, we construct the circuit just as we did in the previous example. The result is Figure 3.

Figure 3

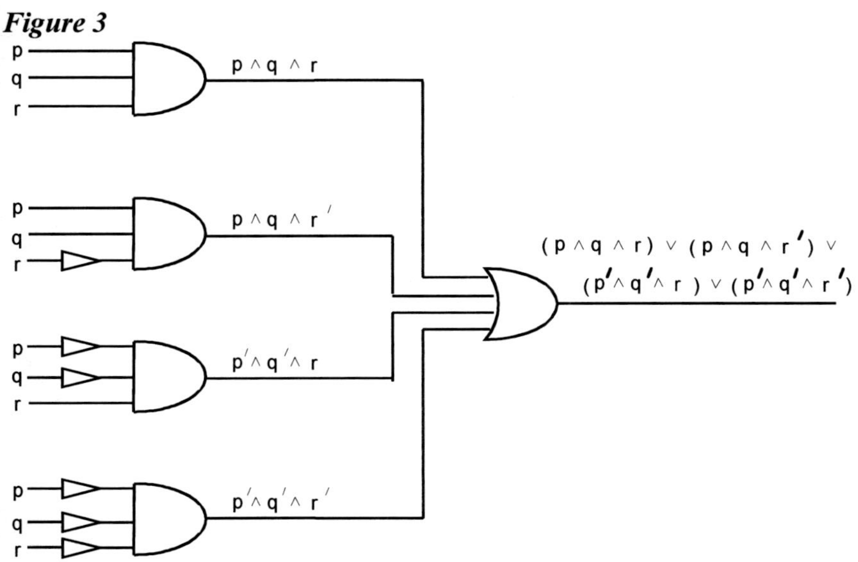

EXERCISES

Define or discuss each of the following terms.

a) AND-OR circuit   b) AND gate
c) OR gate   d) NOT gate
e) Inverter
f) Statement associated with a circuit
g) Truth table of a circuit
h) Comparator

In Exercises 1-5, label the given circuit, determine the statement associated with the circuit, and find the truth table of the circuit.

1.

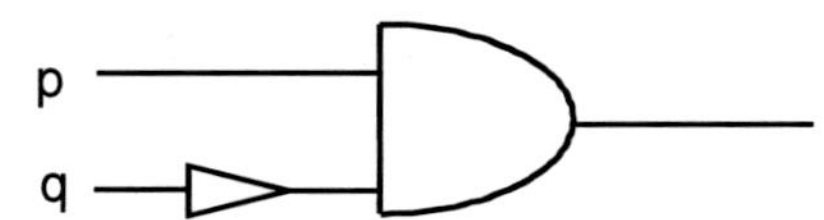

2.

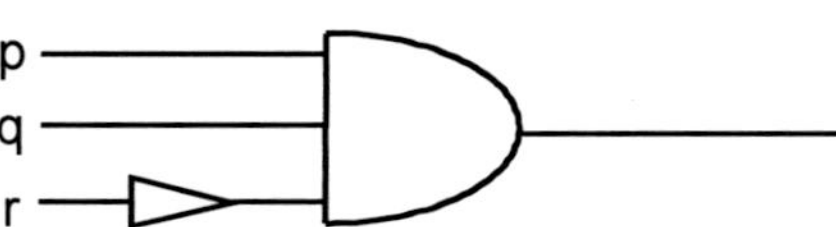

3.

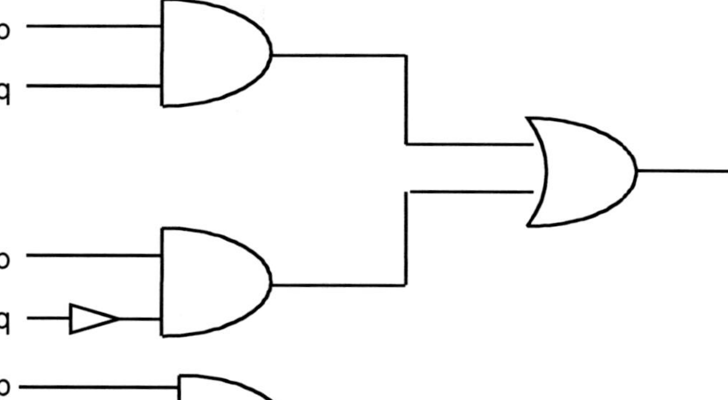

4.

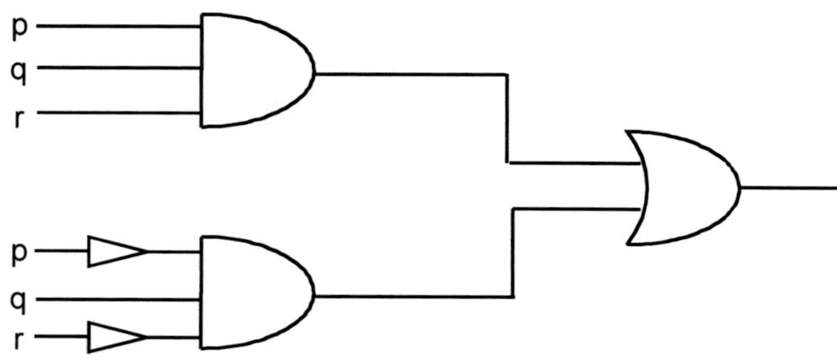

5.

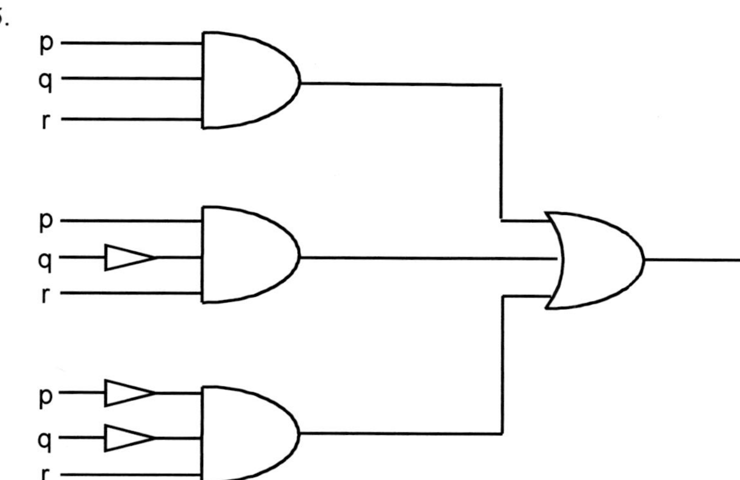

In Exercises 6-15, find the circuit with the given truth table.

6.

p	
T	F
F	T

7.

p	q	
T	T	F
T	F	T
F	T	F
F	F	F

8.

p	q	
T	T	T
T	F	T
F	T	F
F	F	F

9.

p	q	
T	T	F
T	F	T
F	T	F
F	F	T

10.

p	q	
T	T	F
T	F	T
F	T	T
F	F	T

11.

p	q	r	
T	T	T	F
T	T	F	F
T	F	T	F
T	F	F	F
F	T	T	F
F	T	F	F
F	F	T	T
F	F	F	F

12.

p	q	r	
T	T	T	T
T	T	F	F
T	F	T	T
T	F	F	F
F	T	T	F
F	T	F	F
F	F	T	F
F	F	F	F

13.

p	q	r	
T	T	T	F
T	T	F	F
T	F	T	F
T	F	F	T
F	T	T	F
F	T	F	F
F	F	T	F
F	F	F	T

14.

p	q	r	
T	T	T	F
T	T	F	T
T	F	T	F
T	F	F	F
F	T	T	T
F	T	F	F
F	F	T	T
F	F	F	F

15.

p	q	r	
T	T	T	F
T	T	F	T
T	F	T	T
T	F	F	F
F	T	T	T
F	T	F	F
F	F	T	T
F	F	F	F

16. Design a circuit with two inputs p and q that outputs a high voltage if and only if the inputs are the same. (This is a *comparator* with two inputs.)

17. Design a circuit that has three inputs p, q and r, and outputs a high voltage if and only if $p = q$ and $p \neq r$.

18. Design a circuit that has three inputs p, q and r, and outputs a high voltage if and only if $p \neq q$ and $p \neq r$.

19. Design a circuit that has three inputs p, q and r, and outputs a high voltage if and only if *exactly* two of the inputs are the same.

20. A room has two switches that control an overhead light. Whatever the status of the light, flipping either switch changes that status. Design a circuit to make the switches behave in this manner.

21. A certain light is controlled by three switches, labeled p, q and r. Switch p is an *enabling switch*, in that if p is on, the other two switches control the light as in the previous exercise. But if switch p is off, then switch q is deactivated, and only switch r controls the light. Design a circuit with this behavior.

Answers To Odd Numbered Exercises

Section 1.1
3. Simple statement
5. Compound statement
7. Compound statement
9. Not a statement
11. Compound statement
13. No horse can run. (Or: All horses cannot run.)
15. There is a flower in the garden.
17. No cats have kittens.
19. $p \wedge q$
21. $q \wedge \sim p$
23. $p \leftrightarrow q$
25. $\sim p \wedge \sim q$
27. $p \rightarrow q$
29. The ship is sinking or help is not coming.
31. If help is coming, then the ship is not sinking.
33. It is not true that the ship is not sinking.
35. Not all birds are animals. (Or: Some birds are not animals.)
37. No animal is a bird.
39. All birds are animals, and some birds are animals, and no animals are birds.
41. $p \rightarrow q$; $p =$ wishes were dollars, $q =$ I'd be a millionaire.
43. $p \leftrightarrow q$; $p =$ I will go to school, $q =$ today is Friday.
45. $\sim p \rightarrow \sim q$; $p =$ you study, $q =$ you can go to the movies.
47. $p \vee \sim p$; $p =$ to be.
49. $p \rightarrow (q \wedge r)$; $p =$ you eat that chocolate sundae, $q =$ you will get sick, $r =$ you will get fat.

Section 1.2
(The last column in each table is the column corresponding to the given statement.)

1.

p	
T	T
F	F

3.

p	
T	F
F	T

5.

p	q	
T	T	T
T	F	T
F	T	F
F	F	F

7.

p	
T	T
F	T

9.

p	q	
T	T	T
T	F	T
F	T	T
F	F	T

11.

p	q	r	
T	T	T	T
T	T	F	F
T	F	T	F
T	F	F	F
F	T	T	F
F	T	F	F
F	F	T	F
F	F	F	T

13.

p	q	r	
T	T	T	T
T	T	F	T
T	F	T	T
T	F	F	T
F	T	T	T
F	T	F	F
F	F	T	T
F	F	F	T

15.

p	q	r	
T	T	T	T
T	T	F	F
T	F	T	T
T	F	F	F
F	T	T	T
F	T	F	F
F	F	T	T
F	F	F	T

17.

p	q	r	
T	T	T	F
T	T	F	F
T	F	T	F
T	F	F	T
F	T	T	T
F	T	F	T
F	F	T	T
F	F	F	T

19.

p	q	r	s	
T	T	T	T	T
T	T	T	F	T
T	T	F	T	T
T	T	F	F	T
T	F	T	T	T
T	F	T	F	F
T	F	F	T	F
T	F	F	F	F
F	T	T	T	T
F	T	T	F	F
F	T	F	T	F
F	T	F	F	F
F	F	T	T	T
F	F	T	F	F
F	F	F	T	F
F	F	F	F	F

21.

p	q	r	s	
T	T	T	T	F
T	T	T	F	F
T	T	F	T	F
T	T	F	F	F
T	F	T	T	F
T	F	T	F	F
T	F	F	T	T
T	F	F	F	F
F	T	T	T	F
F	T	T	F	F
F	T	F	T	F
F	T	F	F	F
F	F	T	T	F
F	F	T	F	F
F	F	F	T	F
F	F	F	F	F

31.

p	q	p XOR q
T	T	F
T	F	T
F	T	T
F	F	F

33. It is a tautology.

Section 1.3

9. a) Roses are not red or violets are not blue.
    b) Roses are not red and violets are not blue.
11. a) Elephants are not large or flies are large.
    b) Elephants are not large and flies are large.
13. a) Today is not Friday or not Saturday.
    b) Today is not Friday and not Saturday.
15. a) It is either Friday or Saturday.
    b) It is Friday and Saturday.

17. converse:$q \rightarrow\ \sim p$
    inverse:$p \rightarrow\ \sim q$
    contrapositive: $\sim q \rightarrow p$
19. converse:$r \rightarrow (p \wedge q)$
    inverse: $\sim (p \wedge q) \rightarrow\ \sim r$
    contrapositive: $\sim r \rightarrow\ \sim (p \wedge q)$
21. converse:$(r \rightarrow s) \rightarrow (p \rightarrow q)$
    inverse: $\sim (p \rightarrow q) \rightarrow\ \sim (r \rightarrow s)$
    contrapositive: $\sim (r \rightarrow s) \rightarrow\ \sim (p \rightarrow q)$
23. converse: If Abraham Lincoln will be elected president, then he must be alive.
    inverse: If Abraham Lincoln is not alive today, then he would not be elected president.
    contrapositive: If Abraham Lincoln will not be elected president, then he is not alive today.
25. converse: If I have to go to school, then today is not a holiday.
    inverse: If today is a holiday, then I do not have to go to school.
    contrapositive: If I do not have to go to school, then today is a holiday.
27. converse: If I am not going to school, then today is Monday and it is raining.
    inverse: If today is not Monday or it is not raining, then I am going to school.
    contrapositive: If I am going to school, then today is not Monday or it is not raining.
29. a) Whenever there is a T in the last column of a particular row of the truth table for A, there is also a T in the same row of the truth table for B.

Section 2.1

1. Valid	3. Valid	5. Valid
7. Not valid	9. Valid	11. Valid
13. Valid	15. Valid	

17. p is "it is raining", q is "it is snowing"

$p \vee\ \sim q$ invalid argument

$\sim p$

___

q

19. p is "I have math class", q is "today is Friday", r is "I get the car"

$$p \leftrightarrow q \qquad \text{valid argument}$$
$$r \leftrightarrow q$$

$$\overline{}$$

$$r \leftrightarrow p$$

21. p is "we have *glasnost*", q is "democracy is safe", r is "Gorbachev stays in office"

$$p \rightarrow q \qquad \text{invalid argument}$$
$$r \rightarrow p$$

$$\overline{}$$

$$q$$

23. p is "it snows today", q is "the newspaper is right", r is "the radio is right"

$$p \rightarrow (q \vee \sim r) \qquad \text{valid argument}$$
$$p \wedge \sim q$$

$$\overline{}$$

$$\sim r$$

25. p is "is an ostrich", q is "is 9 feet high", r is "in this aviary", s is "belongs to me", t is "lives on mince−pie"

$$q \rightarrow p \qquad \text{valid argument}$$
$$r \rightarrow s$$
$$p \rightarrow \sim t$$
$$\sim q \rightarrow \sim s$$

$$\overline{}$$

$$r \rightarrow \sim t$$

27. p is "is sane", q is "can do logic", r is "fit to serve on a jury", s is "your sons"

$$p \rightarrow q$$
$$\sim p \rightarrow \sim r$$
$$s \rightarrow \sim q$$

This is equivalent to

$$s \rightarrow \sim q$$
$$\sim q \rightarrow \sim p$$
$$\sim p \rightarrow \sim r$$

Hence, the conclusion is $s \rightarrow \sim r$, or "Your sons are not fit to serve on a jury"

29. p is "is a pudding", q is "is nice", r is "this dish", s is "is wholesome"

$$p \rightarrow q$$
$$r \rightarrow p$$
$$q \rightarrow \sim s$$

This is equivalent to

$$r \rightarrow p$$
$$p \rightarrow q$$
$$q \rightarrow \sim s$$

Hence, the conclusion is $r \rightarrow \sim s$, or "This dish is not wholesome"

Section 2.2

1. $p \wedge q'$ 3. $(p \wedge q) \vee (p \wedge q')$

p	q	
T	T	F
T	F	T
F	T	F
F	F	F

p	q	
T	T	T
T	F	T
F	T	F
F	F	F

5. $(p \wedge q \wedge r) \vee (p \wedge q' \wedge r) \vee (p' \wedge q' \wedge r)$

p	q	r	
T	T	T	T
T	T	F	F
T	F	T	T
T	F	F	F
F	T	T	F
F	T	F	F
F	F	T	T
F	F	F	F

7.

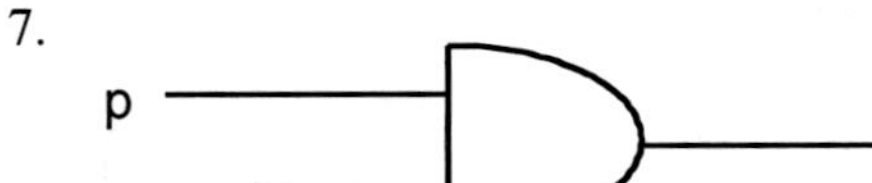

9.

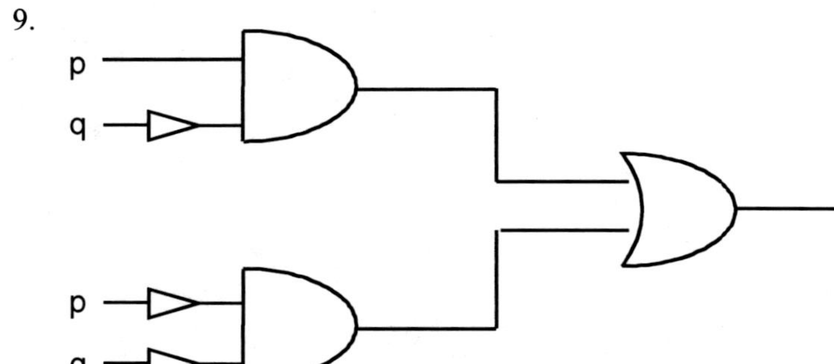

11.

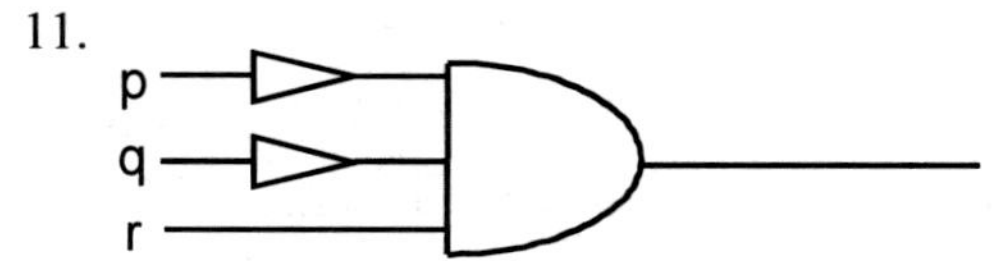

13.

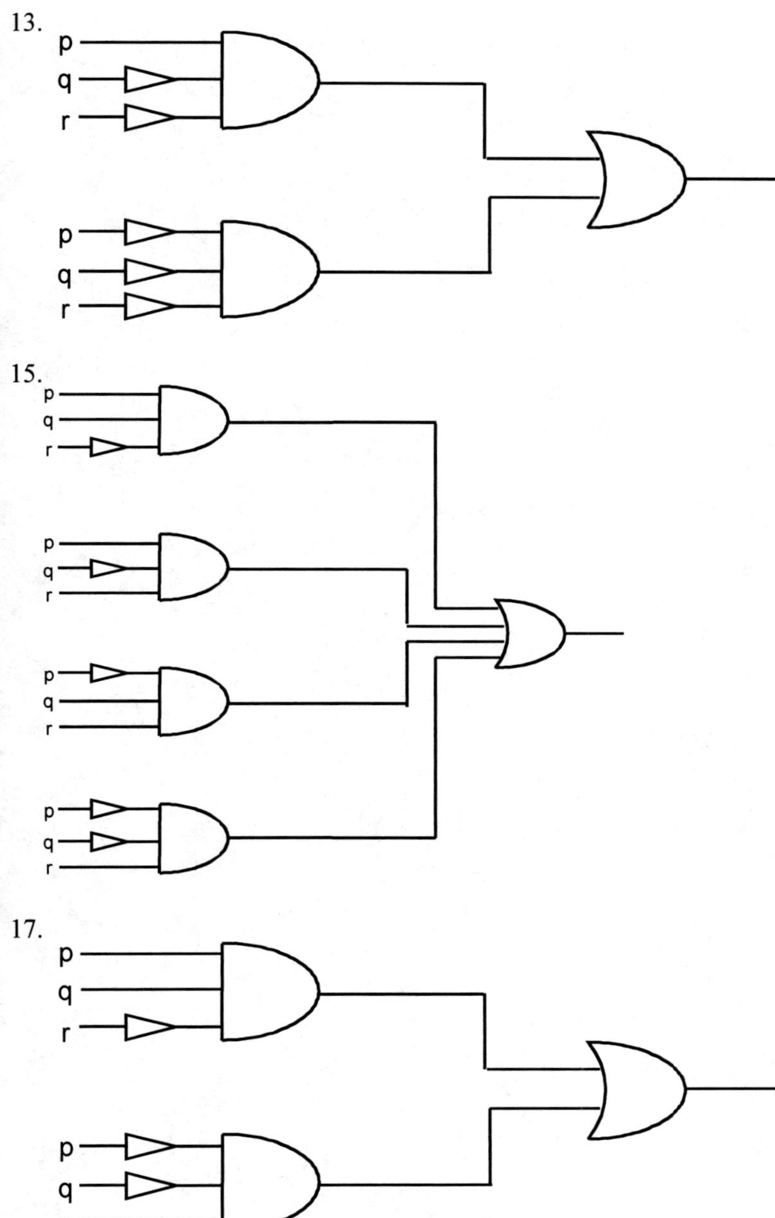

15.

17.

19.

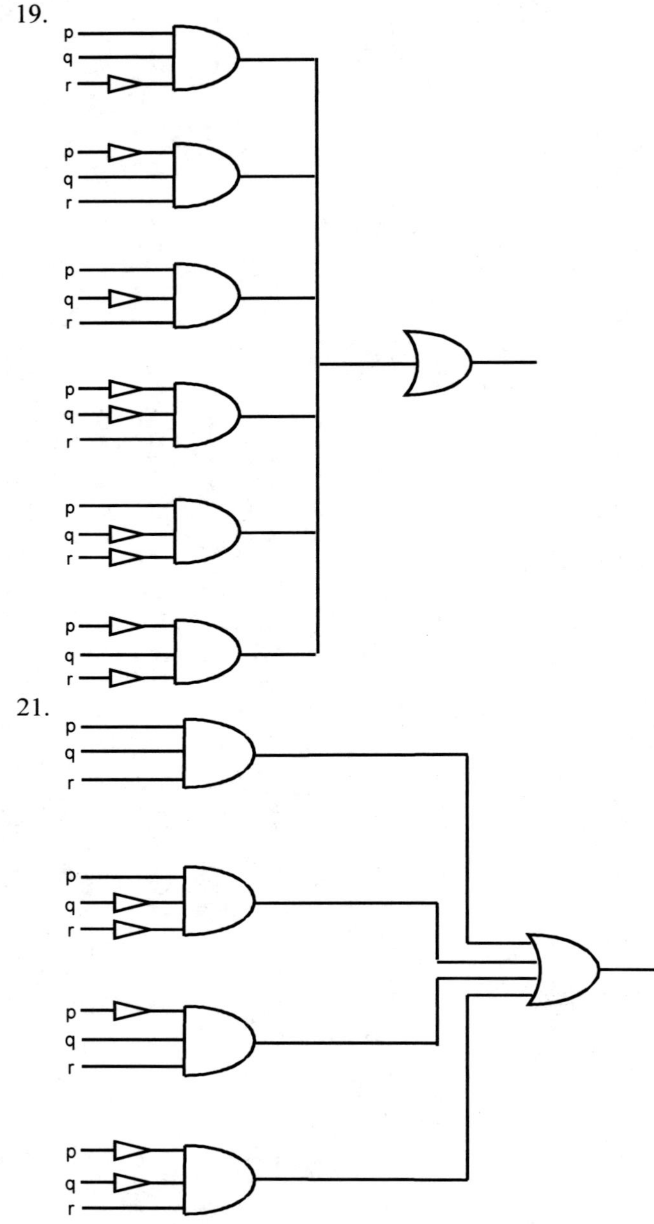

21.